D1393708

LITTLE BOOK OF SHEEP

WEIDENFELD & NICOLSON
LONDON

SPRINGTIME, FEEDING THE LAMBS *Frederick Morgan* 1856-1927

YOUNG LAMBS

The spring is coming by a many signs;
The trays are up, the hedges broken down
That fenced the haystack, and the
remnant shines
Like some old antique fragment
weathered brown.
And where suns peep, in every
sheltered place,
The little early buttercups unfold
A glittering star or two - till many trace
The edges of the blackthorn
clumps in gold.
And then a little lamb bolts up behind
The hill, and wags his tail to meet the ewe;
And then another, sheltered from the wind,
Lies all his length as dead - and lets me go
Close by, and never stirs, but basking lies,
With legs stretched out as though he
could not rise.

John Clare 1793-1864

MR HEALEY'S SHEEP
W. H. Davis

The Derby Ram

As I was going to Derby
Upon a market day,
I met the finest ram, sir,
That ever was fed on hay.

This ram was fat behind, sir,
This ram was fat before,
This ram was three
yards high, sir,
Indeed he was no more.

The wool upon his back, sir,
Reached up unto the sky,
The eagles built their
nests there,
For I heard the young ones cry.

The wool upon his tail, sir,
Was three yards and an ell,
Of it they made a rope, sir,
To pull the parish bell.

The space between the
horns, sir,
Was as far as man could reach,
And there they built a pulpit,
But no one in it preached.

This ram had four legs to
walk upon,
This ram had four legs to stand,
And every leg he had, sir,
Stood on an acre of land.

Now the man that fed the
ram, sir,
He fed him twice a day,
And each time that he fed
him, sir,
He ate a rick of hay.

The man that killed the
ram, sir,
Was up to his knees in blood,
And the boy that held the
pail, sir,
Was carried away in the flood.

Indeed, sir, it's the truth, sir,
For I never was taught to lie,
And if you go to Derby, sir,
You may eat a bit of the pie.

TRADITIONAL

THE SHEEP

LAZY SHEEP, pray tell me why
In the grassy fields you lie,
Eating grass and daisies white,
From the morning till the night?
Everything can something do,
But what kind of use are you?

Nay, my little master, nay,
Do not serve me so, I pray;
Don't you see the wool that grows
On my back to make you clothes?
Cold, and very cold you'd get,
If I did not give you it.

Sure it seems a pleasant thing
To nip the daisies in the spring,
But many chilly nights I pass
On the cold and dewy grass,
Or pick a scanty dinner where
All the common's brown and bare.

Then the farmer comes at last,
When the merry spring is past,
And cuts my woolly coat away
To warm you in the winter's day;
Little master, this is why
In the grassy fields I lie.

Ann Taylor 1782-1866

IN THE VALLEY OF CORWEN, NORTH WALES

R. J. Hammond 19th century

ORPHANS
Charles Edward Wilson 1891-1941

First Sight

LAMBS that learn to walk in snow
When their bleating clouds the air
Meet a vast unwelcome, know
Nothing but a sunless glare.
Newly stumbling to and fro
All they find, outside the fold,
Is a wretched width of cold.

As they wait beside the ewe,
Her fleeces wetly caked, there lies
Hidden round them, waiting too,
Earth's immeasurable surprise.
They could not grasp it if they knew,
What so soon will wake and grow
Utterly unlike the snow.

Philip Larkin 1922-1985

A Sheep Fair

The day arrives of the autumn fair,
And torrents fall,
Though sheep in throngs are gathered there,
Ten thousand all,
Sodden, with hurdles round them reared:
And, lot by lot, the pens are cleared,
And the auctioneer wrings out his beard,
And wipes his book, bedrenched and smeared,
And rakes the rain from his face with the edge
of his hand,
As torrents fall.

The wool of the ewes is like a sponge
With the daylong rain:
Jammed tight, to turn, or lie, or lunge,
They strive in vain.
Their horns are soft as finger nails
Their shepherds reek against the rails,
The tied dogs soak with tucked-in tails,
The buyers' hat-brims fill like pails,
Which spill small cascades when they shift their stand
In the daylong rain.

Thomas Hardy 1840-1928

THE MARKET PLACE, ST. ALBANS

Henry Milbourne 1781-1848

Four Good Legs

THE BIRDS did not understand Snowball's long words, but they accepted his explanation, and all the humbler animals set to work to learn the new maxim by heart. FOUR LEGS GOOD, TWO LEGS BAD, was inscribed on the end wall of the barn, above the Seven Commandments and in bigger letters. When they had once got it by heart, the sheep developed a great liking for this maxim, and often as they lay in the field they would all start bleating 'Four legs good, two legs bad! Four legs good, two legs bad!' and keep it up for hours on end, never growing tired of it.

ANIMAL FARM

George Orwell 1903-1950

SHEEP AND LAMBS

IN AN EXTENSIVE LANDSCAPE

Eugene Verboeckhovn 1799-1881

A Child's Voice

On winter nights shepherd and I
Down to the lambing-shed would go;
Rain round our swinging lamp did fly
Like shining flakes of snow.

There on a nail our lamp we hung,
And O it was beyond belief
To see those ewes lick with hot tongue
The limp wet lambs to life.

A week gone and sun shining warm
It was as good as gold to hear
Those new-born voices round the farm
Cry shivering and clear.

Where was a prouder man than I
Who knew the night those lambs were born,
Watching them leap two feet on high
And stamp the ground in scorn?

Gone sheep and shed and lighted rain
And blue March morning; yet today
A small voice crying brings again
Those lambs leaping at play.

Andrew Young 1855-1971

CHILDREN WITH BABY LAMB
John Lawson early 20th century

FEEDING THE LAMBS

Arthur Trevor Haddon 1864-1941

The Lady of the Lambs

She walks - the lady of my delight -
A shepherdess of sheep.
Her flocks are thoughts. She keeps them white;
She guards them from the steep.
She feeds them on the fragrant height,
And folds them in for sleep.

*

She roams maternal hills and bright,
Dark valleys safe and deep.
Her dreams are innocent at night;
The chastest stars may peep.
She walks - the lady of my delight -
A shepherdess of sheep.

*

She holds her little thoughts in sight,
Though gay they run and leap.
She is so circumspect and right;
She has her soul to keep.
She walks - the lady of my delight -
A shepherdess of sheep.

Alice Meynell 1847-1922

WHO MADE THEE?

Little lamb, who made thee?
Dost thou know who made thee?
Gave thee life, and bid thee feed
By the stream and o'er the mead:
Gave thee clothing of delight,
Softest clothing, woolly, bright;
Gave thee such a tender voice,
Making all the vales rejoice?
Little lamb, who made thee?
Dost thou know who made thee?

THE LAMB *William Blake* 1757-1827

FEEDING THE SHEEP

George Sheridan Knowles 1863-1931

TWO PRIZE BORDER LEICESTER
RAMS IN A LANDSCAPE
Thomas Weaver 1774-1843

Observations On Quadrapeds

AFTER EWES AND LAMBS are shorn, there is great confusion and bleating, neither the dams nor the young being able to distinguish one another as before. This embarrassment seems not so much to arise from the loss of the fleece, which may occasion an alteration in their appearance, as from the defect of that *notus ordo*, discriminating each individual personally; which also is confounded by the strong scent of pitch and tar wherewith they are newly marked; for the brute creation recognize each other more from the smell than the sight; and in matters of identity and diversity appeal much more to their noses than their eyes.

THE NATURAL
HISTORY OF SELBORNE
Gilbert White 1720-1793

SPRING LAMBS *A. W. Redgate* 19th century

Spring Song

On the grassy banks
  Lambkins at their pranks;
Woolly sisters, woolly brothers,
  Jumping off their feet,
While their woolly mothers
  Watch by them and bleat.

Christina Rossetti 1830-1894

A Blushing Ewe

BATHSHEBA, after throwing a glance here, a caution there, and lecturing one of the younger operators who had allowed his last finished sheep to go off among the flock without re-stamping it with her initials, came again to Gabriel, as he put down the luncheon to drag a frightened ewe to his shear-station, flinging it over upon its back with a dexterous twist of the arm. He lopped off the tresses about its head, and opened up the neck and collar, his mistress quietly looking on.

'She blushes at the insult,' murmured Bathsheba, watching the pink flush which arose and overspread the neck and shoulders of the ewe where they were left bare by the clicking shears - a flush which was enviable, for its delicacy, by many queens of coteries, and would have been creditable, for its promptness, to any woman in the world.

FAR FROM THE MADDING CROWD
Thomas Hardy 1840-1928

RURAL SCENE

Gaetano-Stefano 1757-1821

RAMS AND YEWS IN A LANDSCAPE
William Higgins 1820-1884

Baa, Baa, Black Sheep

Baa, baa, black sheep,
Have you any wool?
Yes, sir, yes, sir,
Three bags full;
One for the master,
And one for the dame,
And one for the little boy
Who lives down the lane.

TRADITIONAL

Sheep

When I was once in Baltimore,
A man came up to me and cried,
'Come, I have eighteen hundred sheep,
And we will sail on Tuesday's tide.

'If you will sail with me, young man,
I'll pay you fifty shillings down;
These eighteen hundred sheep I take
From Baltimore to Glasgow town.'

He paid me fifty shillings down,
I sailed with eighteen hundred sheep;
We soon had cleared the harbour's mouth,
We soon were in the salt sea deep.

The first night we were out at sea
Those sheep were quiet in their mind;
The second night they cried with fear -
They smelt no pastures in the wind.

They sniffed, poor things, for their green fields,
They cried so loud I could not sleep:
For fifty thousand shillings down
I would not sail again with sheep.

William Henry Davies 1871-1940

UNLOADING VESSELS
James John Hill 1811-1882

LAMBING TIME
Basil Bradley
1842-1904

The Maiden & The Mountain Lamb

The dew was falling fast, the stars began to blink;
I heard a voice; it said, 'Drink pretty creature, drink!'
And, looking o'er the hedge, before me I espied
A snow-white mountain-lamb with a maiden at its side.

Nor sheep nor kine were near; the lamb was all alone,
And by a slender cord was tethered to a stone;
With one knee on the grass did the little maiden kneel,
While to that mountain-lamb she gave its evening meal.

The lamb, while from her hand he thus his supper took,
Seemed to feast with head and ears; and his tail with pleasure shook.
'Drink, pretty creature, drink,' she said in such a tone
That I almost received her heart into my own.

THE PET LAMB
William Wordsworth 1770-1850

Mary's Lamb

MARY had a little lamb,
    Its fleece was white as snow;
And everywhere that Mary went
    The lamb was sure to go.

It followed her to school one day,
    That was against the rule;
It made the children laugh and play
    To see a lamb at school.

And so the teacher turned it out,
    But still it lingered near,
And waited patiently about
    Till Mary did appear.

Why does the lamb love Mary so?
    The eager children cry;
Why, Mary loves the lamb, you know,
    The teacher did reply.

TRADITIONAL

YOUNG FRIENDS
Lexden Lewis Pocock 1850-1919

Sheep In Winter

The sheep get up and make their many tracks
And bear a load of snow upon their backs,
And gnaw the frozen turnip to the ground
With sharp, quick bite, and then go noising round
The boy that pecks the turnips all the day
And knocks his hands to keep the cold away
And laps his legs in straw to keep them warm
And hides behind the hedges from the storm.
The sheep, as tame as dogs, go where he goes
And try to shake their fleeces from the snows,
Then leave their frozen meal and wander round
The stubble stack that stands beside the ground,
And lie all night and face the drizzling storm
And shun the hovel where they might be warm.

John Clare 1793-1864

SHEEP HERDING,
MID-WINTER
Joseph Farquharson
1846-1935

A SUSSEX SHEEP WASHING

Charles Edward Johnson 1832-1913

The Sheep Dip

If verdant elder spreads
Her silver flowers; if humble daisies yield
To yellow crow-foot, and luxuriant grass,
Gay shearing-time approaches. First, however,
Drive to the double fold, upon the brim
Of a clear river, gently drive the flock,
And plunge them one by one into the flood:
Plung'd in the flood, not long the struggler sinks,
With his white flakes that glisten through the tide;
The sturdy rustic, in the middle wave,
Awaits to seize him rising: one arm bears
His lifted head above the limpid stream,
While the full clammy fleece the other laves
Around, laborious, with repeated toil;
And then resigns him to the sunny bank,
Where, bleating loud, he shakes his dripping locks.

THE FLEECE
John Dyer 1699-1758

Counting Sheep

HALF-AWAKE I walked
A dimly-seen sweet hawthorn lane
Until sleep came;
I lingered at a gate and talked
A little with a lonely lamb.
He told me of the great still night,
Of calm starlight,
And of the lady moon, who'd stoop
For a kiss sometimes;
Of grass as soft as sleep, or rhymes
The tired flowers sang:
The ageless April tales
Of how, when sheep grew old,
As their faith told,
They went without a pang
To far green fields, where fall
Perpetual streams that call
To deathless nightingales.

And then I saw, hard by,
A shepherd lad with shining eyes,
And round him, gathered one by one
Countless sheep, snow-white;
More and more they crowded
With tender cries,
Till all the field was full
Of voices and of coming sheep.
Countless they came, and I
Watched, until deep
As dream-fields lie
I was asleep. *William Kerr*

STUDY OF SHEEP IN A LANDSCAPE
Richard Whitford 19th century

Sheep Stealing

The mountain sheep are sweeter,
But the valley sheep are fatter;
We therefore deemed it meeter
To carry off the latter.

THE MISFORTUNES OF ELPHIN

Thomas Love Peacock 1785-1866

SHEEP RESTING
Thomas Sidney Cooper 1803-1902

THE SHEPHERD'S DAUGHTER

William Kay Blacklock early 20th century

Little Bo-Peep

LITTLE BO-PEEP has lost her sheep,
And can't tell where to find them;
Leave them alone, and they'll
come home,
And bring their tails behind them.

Little Bo-peep fell fast asleep,
And dreamt she heard them bleating;
But when she awoke, she found it a joke,
For they were still a-fleeting.

Then up she took her little crook,
Determined for to find them;
She found them indeed, but it made her
heart bleed,
For they'd all left their tails behind them.

TRADITIONAL

The Sheep Shop

ALICE rubbed her eyes and looked again. She couldn't make out what had happened at all. Was she in a shop? And was that really - was it really a *sheep* that was sitting on the other side of the counter? Rub as she would, she could make nothing more of it: she was in a little dark shop, leaning with her elbows on the counter, and opposite to her was an old sheep, sitting in an armchair, knitting, and every now and then leaving off to look at her through a great pair of spectacles.

'What is it you want to buy?' the sheep said at last, looking up for a moment from her knitting.

'I don't *quite* know yet,' Alice said very gently. 'I should like to look all round me first, if I might.'

'You may look in front of you, and on both sides, if you like,' said the sheep; 'but you can't look all round you - unless you've got eyes at the back of your head.'

THROUGH THE LOOKING GLASS

Lewis Carroll 1832-1898

A FRIENDLY GREETING

Ernest Walbourn early 20th century

THE SHEPHERD'S REST

George Vicat Cole 1833-1893

The Shepherd's Life

GIVES not the hawthorn bush a sweeter shade
To shepherds, looking on their silly sheep,
Than doth a rich embroidered canopy
To kings that fear their subjects' treachery?
O, yes, it doth; a thousand-fold it doth.
And to conclude - the shepherd's homely curds,
His cold, thin drink out of his leather bottle,
His wonted sleep under a fresh tree's shade,
All which secure and sweetly he enjoys,
Is far beyond a prince's delicates,
His viands sparkling in a golden cup,
His body couched in curious bed,
When care, mistrust, and treason wait on him.

KING HENRY VI, PART III
William Shakespeare 1564-1616

An Annual Event

Every year, for two hundred years at least, lambs ran the same race in Whitewell field. In other fields they had their odd games, but here it was always the same.

By the side of one of the paths stood the oak tree, with the seat under it, and a short distance away stood the great spreading ash. The lambs formed up in a line at the oak, and at some signal they raced to the ash, as fast as their tiny legs would go; then they wheeled round and tore back again. They held a little talk, a consultation, with nose-rubbing, friendly pushes, and then off they went again on their race-track.

THE COUNTRY CHILD

Alison Uttley 1884-1976

BLOSSOM TIME
Benjamin Sigmund 19th century

BRINGING HOME THE SHEEP

Ernest Walbourn early 20th century

A Dog's Life

✵

Lamb led an unusual life for a sheep. She was favoured as if she were a dog. She could come in and out of the cottage as she wished, and when she was in the mood she would join Eva and Acid the dogs, Polly the parrot, Sim and Val the cats, for a share of the food at meal-times. She was part of the household. And at night she either slept in the garden, in a small hut when the weather was bad, or in the grass field on the other side of the garden wall . . .

She had been looked after and loved by Jane's family for over five years. They were her life. I used to pass by their cottage and see her lying in the doorway, reminding me of a Newfoundland dog.

A DRAKE AT THE DOOR
Derek Tangye

SHEEP

There have been times I have been a sheep
happy
chewing cauliflower with various good friends
in a sloping field
  when an old farmer in gumboots
his cap worn backwards, unlatches the
top gate with a bustling black-and-white
dog - and we
  bleat complaints about this
and, pretending we are doing it for
reasons of our own, we huddle into a
small, slow, shamefaced group
  and trot away
back where we should have been.

Alan Brownjohn

A BEAUTIFUL SUMMER'S DAY
James John Hill 1811-1882

SPRINGTIME *Luigi Chialiva* 1842-1914

Little Boy Blue

Little Boy Blue,
Come blow your horn!
The sheep's in the meadow,
The cow's in the corn.

But where is the boy
Who looks after the sheep?
He's under a haycock,
Fast asleep!

Will you wake him?
No, not I
For if I do,
He's sure to cry.

TRADITIONAL

Acknowledgements

Copyright © Weidenfeld and Nicolson 1993
First published in Great Britain in 1993 by
George Weidenfeld and Nicolson Ltd
Orion House, 5 Upper St Martin's Lane,
London WC2H 9EA

All rights reserved. No part of this publication may be reproduced, stored in a retrieval system, or transmitted in any form or by any means, electronic, mechanical, photocopying or otherwise, without the prior permission in writing of the copyright owners.

British Library Cataloguing in Publication Data. A catalogue record for this book is available from the British Library.

Designed and edited by
THE BRIDGEWATER BOOK COMPANY
Words and Pictures chosen by
RHODA NOTTRIDGE
Typesetting by VANESSA GOOD
Printed in Italy

The publishers wish to thank the following for the use of pictures:
THE BRIDGEMAN ART LIBRARY: front cover and pages 20,25,26,39. E.T. ARCHIVE: page 4. FINE ART PHOTOGRAPHS: back cover and pages 2, 7, 8, 11, 13, 15, 16, 19, 22, 29, 30, 33, 35, 36, 41, 42, 43, 45, 46, 49, 50, 53, 54.

The publishers gratefully acknowledge permission to reproduce the following material in this book:
p.9 *First Sight* by Philip Larkin from Whitsun Weddings by permission of F
aber and Faber Ltd.
p.12 *Animal Farm* by George Orwell by permission of the late Sonia Brownell Orwell and Martin Secker & Warburg Ltd and Harcourt, Brace, Jovanovich, USA.
p.14 *A Child's Voice* by Andrew Young by permission The Andrew Young Estate.
p.28 *Sheep* by W.H.Davies; The Estate of W.H.Davies and Jonathan Cape.
p.42 *The Country Child* by Alison Uttley by permission of Faber and Faber Ltd.
p.51 *A Drake At The Door* by Derek Tangye (Michael Joseph, 1972) copyright
© Derek Tangye,
1972 by permission of Michael Joseph and Laurence Pollinger Ltd (USA).
p.52 *Sheep* by Alan Brownjohn by permission the author, © Alan Brownjohn.

Every effort has been made to trace all copyright holders and obtain permissions. The editor and publishers sincerely apologise for any inadvertent errors or omissions and will be happy to correct them in any future edition.

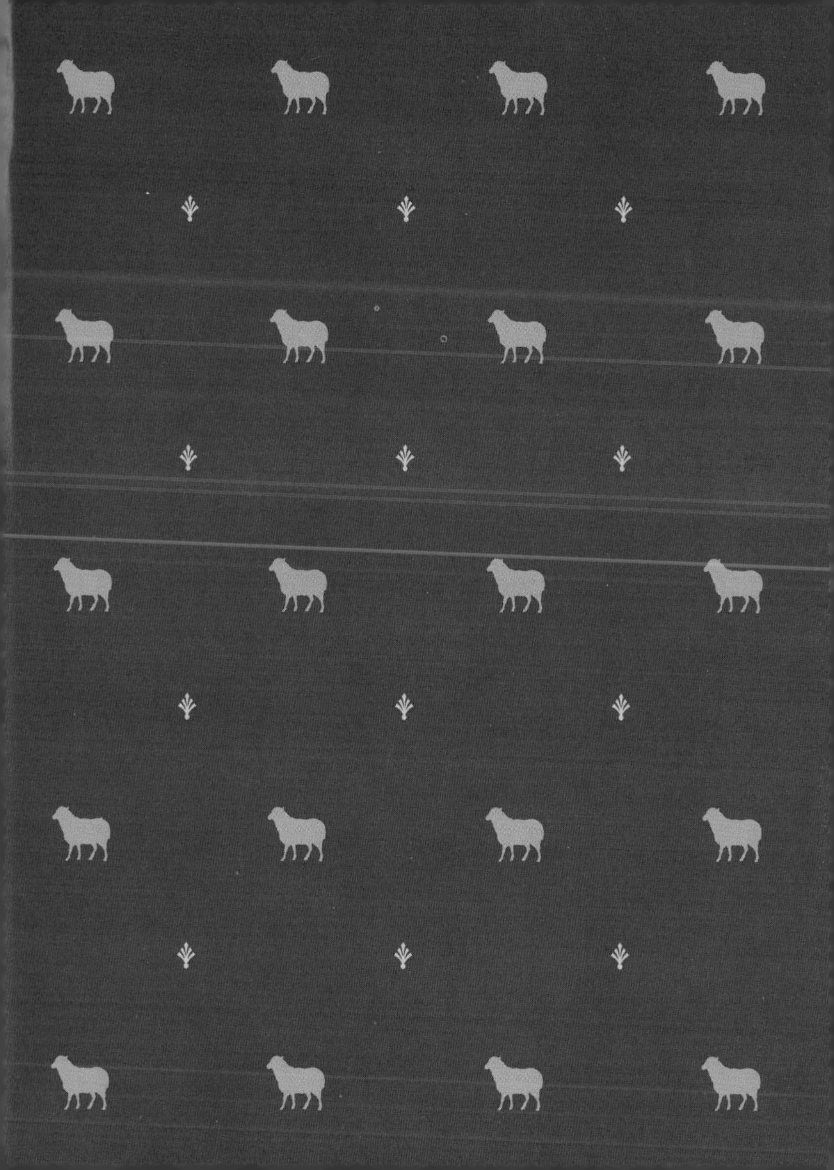